An Essay on the Nature of Space-time

Including Universal Expansion and Dark Energy

by

Dennis Morris

Published by: Abane & Right

31/32 Long Row

Port Mulgrave

Saltburn

TS13 5LF

United Kingdom

01287 678918

dennis355@btinternet.com

May 2016

Contents

Contents

Introduction

This small volume presents a view of empty space that is different from any view that has before been voiced. The view presented here is that our unique 4-dimensional space-time emerges from the $C_2 \times C_2$ finite group as an emergent expectation space. It is utterly remarkable that from all the infinitude of finite groups only the $C_2 \times C_2$ finite group has an emergent expectation space which is manifest as a geometric space. It is quite awesome that this solitary emergent expectation space is our 4-dimensional space-time.

As far as is possible, the detailed mathematics which substantiates this view has been avoided in this volume and only simple mathematics has been used in this presentation. The detailed mathematics that underlies the view of space presented here is presented in detail in other books by the same author.

Unfortunately, it has not proved possible to avoid mathematical terms completely. In particular, the short chapter on gauge theory, which is the final chapter, cannot be presented without reference to covariant differentiation and affine connection. It is not necessary to completely understand the chapter on gauge theory to understand the view of space presented in this essay, and the reader can dispense with that chapter if the mathematical terms discomfort her. This chapter is included for completeness.

I have endeavoured to produce a short and inexpensive volume in the hope that it will find a large readership. The volume has been kept short because I hope to make it easily digestible. The volume is not as inexpensive as I would have wished, and I apologise to the reader for providing so few pages for the cost, but I hope the reader will feel that the content of these few pages is worth that cost.

The view of space presented in this volume is unique in that for the first time we have an understanding of the 4-dimensional space-time of our universe which is more than only awe and mystery. For the first time, we have been able to deduce the properties, the dimension, the rotations and the distance function, of the space-time of the universe. We believe we also now understand why the universe is expanding and why it had an inflationary beginning.

Previous to this volume, no one had any idea why empty space-time exists or why it is 4-dimensional. No one knew why our 4-dimensional space-time has the quadratic distance function it does have or how empty space could expand as our universe clearly is doing. It might be that the view presented in this volume will be modified as our understanding of empty space matures, but we think we now have the essence of the correct understanding.

As well as understanding our 4-dimensional space-time, we also understand every spinor space of all dimensions. If you want an 11-dimensional space, it will have to be a spinor space, but we can tell you exactly what is available – there are eleven different types of 11-dimensional spinor spaces. There are 125 different 15-dimensional spinor spaces. You pick a dimension; we tell you what spaces are available. If you choose 4-dimensions, you get lucky because this is the only number of dimensions that holds anything other than spinor spaces; it holds both our 4-dimensional space-time and the quaternion emergent space as well as a set of spinor spaces. Four dimensions is very special.

We hope the reader will find this volume easy to read. We have deliberately kept it short. If the reader wishes to know more of these matters, there are other more mathematically detailed books which are listed at the end of this volume.

Chapter 1

An Overview of our 4-dimensional Space-time

We take our 4-dimensional empty space to be as we observe it. By observe, we mean detailed examination; we do not mean as seen by an inebriated one-eyed individual with myopia on a galloping horse passing through our space-time on a foggy moonless night.

Newton's view of space and time:

The reputedly sober and the perspicacious Isaac Newton (1642-1726), while sitting sedately in the early morning of a sunlit day, believed that we live in a 3-dimensional space and in a 1-dimensional time. Newton believed we live in both of two types of space; these were the 3-dimensional space and the 1-dimensional time. Our space-time does appear to be that way until we examine it closely using objects that move at very high velocities.

Newton's view is that we have a 1-dimensional time and that fixed to this 1-dimensional time, at each point of it, there is a 3-dimensional space. Newton made no attempt to explain why our space-time was as he supposed it to be; he simply took it to be obvious meaning 'as observed'.

Newton's view of two or more spaces tied together somehow is what mathematicians call a fibre bundle. The reader might find it strange that we could simultaneously exist in more than one space, but it is not new to the reader; we have become accustomed to this view throughout our lifetimes.

We have a picture of a fibre bundle. In this case, the 'base space' is a 1-dimensional space, like Newton's view of time, and the 'attached space' is a 2-dimensional Euclidean plane.

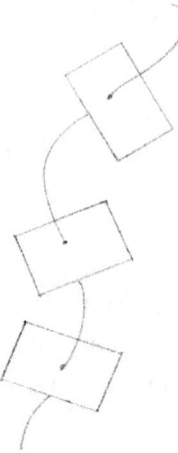

There can be more than one 'attached space' fixed to the 'base space'.

We now know that Newton's view of space and time as separate spaces is wrong. None-the-less, remarkably, it does seem that we do live in two or more separate spaces as Newton had thought but that these separate spaces are our 4-dimensional space-time (the base space) and a few (the number is not yet settled) other spaces called gauge spaces (the attached spaces). These spaces are somehow 'tied together' to form our universe. We will meet the gauge spaces later.

Time is not separate from space:
Detailed examination of our space and time reveals that, as is most easily seen at high velocities, we can rotate in a 2-dimensional space-time plane within our 4-dimensional space-time. The time axis can become transformed into a spatial axis and vice-versa. This is just the theory of special relativity which is widely understood and well known. Indeed, the special theory of relativity is no more than the realisation that we can rotate in a 2-dimensional space-time plane and that physical processes are the same regardless of the direction in which we look (the direction is our velocity) in that 2-dimensional space-time plane[1].

---

[1] A change of velocity is no more than a change of direction in a 2-dimensional space-time plane.

If we can rotate in a 2-dimensional space-time plane, then time must be part of space. If time was a space entirely separate the from 3-dimensional space, then we would not be able to rotate the time axis into the 3-dimensional space. Indeed, the often heard statement that we live in a single 4-dimensional space-time is no more than a statement that we can rotate within a '2-dimensional' space-time 'plane' – well, three such 2-dimensional planes to be precise.

We take it that the reader is familiar with special relativity[2], and so we will not belabour the unity of space and time into one space. We simply state that we live in a single 4-dimensional space with three spatial axes and one temporal axis[3].

Rotations in our 4-dimensional space-time:
Our space-time has the following properties:

1) It is 4-dimensional
2) It has 2-dimensional rotation

The astute reader might be shocked by the apparent conflict between a 4-dimensional space and the 2-dimensional rotations. Why not 4-dimensional rotations in a 4-dimensional space? That would seem more reasonable. Actually, it is more likely that the reader will be shocked by the question being asked than by the facts underlying the question.

We have become accustomed to the idea that all rotations are 2-dimensional; they are not. Only the two 2-dimensional types of rotation are 2-dimensional. The three 3-dimensional types of rotation are 3-dimensional; the six 4-dimensional types of rotation are 4-dimensional. There are rotations of all dimensions. This will be made clear in the later chapter on spinor spaces; for now, we ask the reader to simply accept that there are rotations of dimension higher than two.

---

[2] See: Dennis Morris : Empty Space is Amazing Stuff
[3] Even as your author sits far distant from the reader, your author can hear the 'tick, tick, whirl, whirl' of the reader's brain as she asks, "Why one time axis? Why not two? Why not all spatial axes? Why four axes rather than, say, five or three?

While we are at it, another question. Why do we not see 3-dimensional rotations in our 4-dimensional space-time? Surely, there's room enough. Please note that a 3-dimensional rotation is not a composition of 2-dimensional rotations – see later.

By now, unless the reader is already familiar with the spinor rotations which we will reveal later in this book, the reader will be utterly bewildered – 4-dimensional rotations – never heard of such stuff.

We ask the reader unfamiliar with rotations of dimension higher than two to stay with us. You will find this book most stimulating. We repeat our exposition of shock presented earlier.

It is utterly unreasonable to have 2-dimensional rotations in 4-dimensional space. Surely, the universe must be insane.[4]

The nature of 2-dimensional rotation in our space-time:
We present the six 2-dimensional rotations of our 4-dimensional space-time. There are three space-time rotations:

$$\begin{bmatrix} \cosh\chi & \sinh\chi & 0 & 0 \\ \sinh\chi & \cosh\chi & 0 & 0 \\ 0 & 0 & 1 & 0 \\ 0 & 0 & 0 & 1 \end{bmatrix}, \quad \begin{bmatrix} \cosh\chi & 0 & \sinh\chi & 0 \\ 0 & 1 & 0 & 0 \\ \sinh\chi & 0 & \cosh\chi & 0 \\ 0 & 0 & 0 & 1 \end{bmatrix}$$

$$\begin{bmatrix} \cosh\chi & 0 & 0 & \sinh\chi \\ 0 & 1 & 0 & 0 \\ 0 & 0 & 1 & 0 \\ \sinh\chi & 0 & 0 & \cosh\chi \end{bmatrix} \quad (1.1)$$

and there are three spatial (Euclidean) rotations:

---

[4] Very often, advancement in physics in made by asking a previously unasked, and seemingly insane, question.

$$\begin{bmatrix} 1 & 0 & 0 & 0 \\ 0 & \cos\theta & \sin\theta & 0 \\ 0 & -\sin\theta & \cos\theta & 0 \\ 0 & 0 & 0 & 1 \end{bmatrix}, \begin{bmatrix} 1 & 0 & 0 & 0 \\ 0 & 1 & 0 & 0 \\ 0 & 0 & \cos\theta & \sin\theta \\ 0 & 0 & -\sin\theta & \cos\theta \end{bmatrix}$$

$$\begin{bmatrix} 1 & 0 & 0 & 0 \\ 0 & \cos\theta & 0 & \sin\theta \\ 0 & 0 & 1 & 0 \\ 0 & -\sin\theta & 0 & \cos\theta \end{bmatrix}$$

(1.2)

Note that the arrangement of the rotations could have been swapped. It is only convention that we put the space-time rotations on the top row and down the left-most column.

The eigenvectors of a rotation matrix correspond, roughly, to the axes of the space[5]. If we take the eigenvectors of any of the above six matrices, we find that two of the eigenvectors are independent of the rotation angle – these correspond to the 1's on the leading diagonals of the above rotation matrices, (1.1) & (1.2). For example, for the last rotation matrix in (1.2), two of the four eigenvectors are:

$$\begin{bmatrix} 1 \\ 0 \\ 0 \\ 0 \end{bmatrix} \quad \& \quad \begin{bmatrix} 0 \\ 0 \\ 1 \\ 0 \end{bmatrix}$$

(1.3)

This means these 2-dimensional rotations are rotations about two axes. Rotation about an axis, or, in this case, two axes, is most unusual rotation – it really is – see later. Most rotations are spinor rotations; these are rotations about a single point.

If we had a 4-dimensional rotation within our 4-dimensional space-time, then there would be no eigenvectors which were independent of the rotation angle. This would be rotation about a point and not rotation about an axis. An example of rotation about a point is 2-dimensional

---

[5] This is picturesquely correct and mathematically almost kind of correct.

rotation in the 2-dimensional complex plane, $\mathbb{C}$. Because the complex plane is 2-dimensional, there is no axis 'left over' to be an axis of rotation for the 2-dimensional rotation. The 2-dimensional rotation in the complex plane is:

$$\begin{bmatrix} \cos\theta & \sin\theta \\ -\sin\theta & \cos\theta \end{bmatrix} \tag{1.4}$$

There are two eigenvectors of this matrix, (1.4), and neither of them are independent of the angle of rotation – there are no 1's on the leading diagonal. This, (1.4), is not rotation about an axis while it is within its own space. This, (1.4), 2-dimensional rotation is rotation about an axis, or more than one axis, only when it is within a higher dimensional space as it is in our 4-dimensional space-time, like (1.2).

The reader who has not met spinor rotation before might find this alarming. It is quite simple. The complex plane is 2-dimensional, and the above rotation, (1.4), is rotation in a 2-dimensional plane. There is no other dimension to be an axis. Don't feel left out; it shocks everyone.

More upon our 4-dimensional space-time:
Our 4-dimensional space-time is of the form of four 1-dimensional spaces fitted together, $\mathbb{R}^4$, with six 2-dimensional angles. Where do the four copies of $\mathbb{R}^1$ come from? Where do the six 2-dimensional angles come from, and why are there three of each type of 2-dimensional angle?

Angles exist in only spinor spaces (types of complex numbers). The 2-dimensional Euclidean angle exists in only the 2-dimensional complex plane, $\mathbb{C}$. The 2-dimensional space-time angle exists in only the 2-dimensional hyperbolic complex plane, $\mathbb{S}$, which is 2-dimensional space-time. So what are these angles doing in our 4-dimensional space-time? Clearly, by some means, our 4-dimensional space-time holds within it the two 2-dimensional spinor spaces.

Visualise a horizontally spinning disc. At first, you might think the disc is rotating in only one 2-dimensional plane, but consider a point on the outer rim of that disc; as the disc rotates, the point is accelerating in the two orthogonal directions of the 2-dimensional plane. Such acceleration is rotation in two 2-dimensional space-time planes. The disc is rotating in three 2-dimensional planes.

To understand our 4-dimensional space-time, we need to understand how we get four 1-dimensional spaces held together by six 2-dimensional angles. We can make a start by asking what our space would be like if we had only, say, five angles rather than six.

This is a subtle challenge to the established understanding of space. Traditionally, mathematicians are quite certain in their belief that three copies of the real numbers, $\mathbb{R}^3$, is 3-dimensional space and that four copies of the real numbers, $\mathbb{R}^4$, is 4-dimensional space and that these two spaces are distinct. We are quite certain that a space with six angles is a space distinct from one with five angles. Whereas mathematicians treat angles with disdain, just get as many as you like out of the fridge when you need them, we treat angles as the equals of axes. We say that you need angles as well as axes to make a space, and that spaces of the same dimension but with different numbers of angles are distinct spaces. Ultimately, this 'angles are important' stance explains the directionally quantitised nature of electron spin.

Only five angle space-time:
Imagine a 3-dimensional Euclidean space with only two Euclidean spatial rotation angles. Imagine being able to rotate in both the orthogonal vertical spatial planes but not in the horizontal spatial plane. Our spatial rotations would be limited to two 'hoops' which intersect at the top and the bottom. We have a picture:

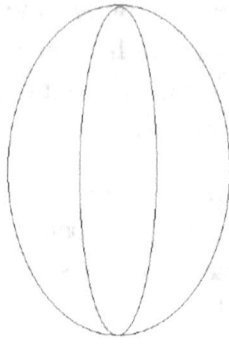

We could not compose these two rotations together to form a rotation which had a horizontal component because that would need a horizontal angle. We could not rotate these 'hoops' around a 'north-south' axis because we have no horizontal angle. Within our 'only five angles' space' spatial rotation would be in only two vertical planes. The axes of these spatial rotations would be fixed as either left or right with nothing between these two directions. Before the reader dismisses this view as fanciful, please be aware that the spin of sub-atomic particles like the electron is exactly of the nature just described. Electrons seem to rotate in a 4-dimensional space that has only two angles; it is 'missing' a few angles.

Suppose, instead of taking away one of the spatial angles of our 4-dimensional space-time, we take away one of the space-time angles. We can now accelerate (change velocity) in an up-down direction or in an east-west direction but not in a north-south direction. Would we be aware of the existence of a north-south direction? Indeed, back to the only two spatial rotations pictured above, in such a space, would we be aware of the horizontal plane?

We refer to our 4-dimensional space-time as being 4-dimensional because we see it as being comprised of four copies of 1-dimensional space; perhaps we should refer to our 4-dimensional space-time as 6-angled space. Why, in our view of space, should we prioritise the number of 1-dimensional spaces over the number of 2-dimensional angles?

We will later meet the quaternion emergent expectation space[6]. This space is a 4-dimensional space in that it has four copies of 1-dimensional space, but it is a 2-angled space in that it has only two 2-dimensional angles. Rotation in this space is limited to the kind of rotation (intrinsic spin) we observe associated with the electron. It seems that the number of angles in a space is at least as important as the number of 1-dimensional axes in that space.

The gentle reader might think that the number of 2-dimensional angles in a space is dictated by the distance function of that space. The 2-dimensional Euclidean rotation respects the Pythagorean distance function:

$$d^2 = x^2 + y^2 \tag{1.5}$$

The 2-dimensional space-time rotation respects the distance function:

$$d^2 = t^2 - z^2 \tag{1.6}$$

Our 4-dimensional space-time has distance function:

$$d^2 = t^2 - x^2 - y^2 - z^2 \tag{1.7}$$

We see that there are six ways we can pair together two of the four variables in the 4-dimensional space-time distance function, (1.7):

$$
\begin{aligned}
d^2 &= t^2 - x^2 & \qquad d^2 &= x^2 + y^2 \\
d^2 &= t^2 - y^2 & \qquad d^2 &= x^2 + z^2 \\
d^2 &= t^2 - z^2 & \qquad d^2 &= y^2 + z^2
\end{aligned}
\tag{1.8}
$$

The reader might think this automatically means there are six 2-dimensional angles in our 4-dimensional space-time. This does mean that our 4-dimensional space-time could support six 2-dimensional angles if there are the six 2-dimensional angles available, but the angles might not be available; perhaps there are none left in the fridge. It seems that there are six 2-dimensional angles available because out 4-

---

[6] Emergent expectation spaces and their emergent expectation distance functions will be explained in a later chapter.

dimensional space-time is the emergent expectation space of the six A_3 spinor algebras. There is one angle in each A_3 algebra; it is because there are six A_3 algebras that there are six angles available to form our 4-dimensional space-time. In the case of the quaternion emergent expectation space, there are only two 4-dimensional quaternion algebras, and so there are only two angles available to form the quaternion emergent expectation space. The quaternion emergent expectation space has the distance function:

$$d^2 = t^2 + x^2 + y^2 + z^2 \qquad (1.9)$$

This is able to support six 2-dimensional Euclidean rotation angles, but, because there are only two such angles available from the two quaternion algebras, this space, (1.9), has rotation in only two 2-dimensional planes. This explains why the axis of the intrinsic spin of the electron is strictly either up or down.

There is something here that perhaps needs more emphasis. For centuries, mathematicians and physicists have taken angles for granted; they have assumed the existence of angles in every type of space they have explored without giving the genesis of these angles a second thought. Mathematicians have taken the number of axes to be the defining property of a space and have blithely ignored any role played by angles[7]. Yet, clearly 1-dimensional axes do not fit themselves together, and there are no angles in 1-dimensional space. Your author is asserting that angles have their own existence independent of axes just as 1-dimensional axes exist independently of angles. A geometric space is formed when these independent objects, angles and axes, come together in the correct numbers. This is a view very different from the standard mantra[8].

So, we are saying that to form a space in which we can freely wave our arms around, we need the proper number of 1-dimensional spaces and

---

[7] The metric space axioms completely disregard angles.
[8] Do not trust your author's opinion. Think for yourself. Form your own opinion.

we need the proper number of angles. It is not sufficient to have only the proper number of 1-dimensional spaces. Wow! This is a new view.

An example with 3-dimensional angles:
There are three types of 3-dimensional angles[9]. One of these angles respects the distance function:

$$d^3 = a^3 + b^3 + c^3 - 3abc \qquad (1.10)$$

It might have been that nature would have a 4-dimensional emergent expectation space with distance function:

$$d^3 = a^3 + b^3 + c^3 + d^3 - 3abc - 3abd - 3acd - 3bcd \qquad (1.11)$$

We see that setting any one of the four variables to zero will give a 3-dimensional distance function of the form above (1.10). Thus, such a space, (1.11), could support four 3-dimensional rotations. If there were four 3-dimensional angles available, we would be able to (3-dimensionally) wave our arms around in such a space. There is no such 4-dimensional space.

The only 3-dimensional emergent expectation space[10] has the distance function:

$$d^3 = a^3 + 3abc \qquad (1.12)$$

There are no rotations, either 2-dimensional or 3-dimensional, which respect this distance function[11], and so there is no rotation in this emergent expectation space. As such, this is not a geometric space in any meaningful sense of the concept of space.

---

[9] See: Dennis Morris : Complex Numbers The Higher Dimensional Forms – 2nd edition.
[10] We repeat, emergent expectation spaces will be described in a later chapter.
[11] A rotation holds the distance from the origin invariant. That distance function is the determinant of the matrix form of the division algebra. That distance is defined by an expression like (1.10). There is no rotation corresponding to the 'distance function' (1.12) because there is not division algebra whose matrix form has a determinant of this form.

What is an angle?:

An angle is a variable, or a function of variables, which appears in a trigonometric function within a rotation matrix.

Rotation matrices derive from the finite groups. We take a finite group of order N written as a set of $N \times N$ permutation matrices[12]; we multiply each permutation matrix by a real number; we add the separate matrices, and we take the matrix exponential of this sum. This produces the polar form of a type of complex numbers. We give an example with the order three finite group C_3:

$$\left\{ \begin{bmatrix} 1 & 0 & 0 \\ 0 & 1 & 0 \\ 0 & 0 & 1 \end{bmatrix}, \begin{bmatrix} 0 & 1 & 0 \\ 0 & 0 & 1 \\ 1 & 0 & 0 \end{bmatrix}, \begin{bmatrix} 0 & 0 & 1 \\ 1 & 0 & 0 \\ 0 & 1 & 0 \end{bmatrix} \right\}$$

$$\exp\left(\begin{bmatrix} a & b & c \\ c & a & b \\ b & c & a \end{bmatrix} \right) = \begin{bmatrix} e^a & 0 & 0 \\ 0 & e^a & 0 \\ 0 & 0 & e^a \end{bmatrix} \begin{bmatrix} v_A(b,c) & v_B(b,c) & v_C(b,c) \\ v_C(b,c) & v_A(b,c) & v_B(b,c) \\ v_B(b,c) & v_C(b,c) & v_A(b,c) \end{bmatrix}$$

$$(1.13)$$

The rotation matrix is the right-most matrix with the $v_i(b,c)$ functions within it. The $v_i(b,c)$ functions are the trigonometric functions of the particular spinor algebra; the (b,c) is the angle. Note that an angle might contain, and usually does contain, more than one variable. Only the 2-dimensional trigonometric functions have a single variable angle. In general, a N-dimensional trigonometric function will have $(N-1)$ variables as an angle. The variables within the angle are 'tied together' in the same way that the elements of the underlying group are 'tied together'.

---

[12] A permutation matrix is a square matrix with a single 1 in each row and a single 1 in each column and zeros everywhere else.

The uniqueness of our 4-dimensional space-time:

We have seen above that there are many parts of a space which need to come together to form a geometric space in which we can freely wave our arms around as we can in our 4-dimensional space-time. We need:

1) The right number of angles available to suit the number of 1-dimensional spaces.
2) A number of 1-dimensional spaces.
3) A distance function which, when the appropriate number of its variables are zero, will reduce to a distance function respected by the lesser dimensional rotations. We need this to happen for all sets of appropriate numbers of variables.

Remarkably, of all the emergent expectation spaces of all the infinite number of finite groups, the only emergent expectation space with all these required properties is the emergent expectation space of the A_3 algebras which is our 4-dimensional space-time. Our 4-dimensional space-time is unique[13].

That the only expectation space to emerge from all the finite groups exactly matches our 4-dimensional space-time is compelling evidence that the mathematical technique of forming an emergent expectation space is a correct way to proceed.

---

[13] See: Dennis Morris : The Uniqueness of our Space-time.

Chapter 2

A Look at Spinor Spaces

The real numbers are derived from the existence of the number one. This was shown by Bertrand Russell (1872-1970) circa 1900. The finite groups are no more than closed sets of permutations. The objects being permuted can be anything, and so we could use different real numbers. For example, the group C_3 can be seen as the three ordered permutations of $\{1,2,3\}$ which are $\{123, 231, 312\}$. We have above, (1.13), given an example of how one of the four types of 3-dimensional complex numbers is derived from the group C_3.[14] All four types of 3-dimensional complex numbers derive from the group C_3. There are no types of complex numbers, also called division algebras and also called spinor algebras or spinor spaces, which are not derived from the finite groups; every division algebra has a finite group at its core.

The spaces associated with types of complex numbers are spinor spaces. A spinor space has a single real number axis and a number of imaginary axes. The dimension of a spinor space is the total number of axes, of course. One example of a spinor space is the complex plane, $\mathbb{C}$; another example is either of the two quaternion spaces. A spinor space is just a division algebra, and there are exactly as many spinor spaces as there are division algebras.

Sometimes we see these spinor spaces written as vectors of complex numbers; for example, a Weyl spinor is an ordered pair of complex numbers. This is just another, somewhat obscure, way of writing a quaternion. A Dirac spinor is just a pair of quaternions.

---

[14] More details can be found in Dennis Morris : Complex Numbers The Higher Dimensional Forms or in Dennis Morris : The Physics of Empty Space.

The distance function of a spinor space:

The nature of a single spinor space is that it has a distance function. The distance function is the determinant of the matrix of which we take the exponential to form the polar form of the algebra. For example, the distance function of the 3-dimensional spinor space we have given as an example above, (1.13), is:

$$\det\left(\begin{bmatrix} a & b & c \\ c & a & b \\ b & c & a \end{bmatrix}\right) = a^3 + b^3 + c^3 - 3abc \qquad (2.1)$$

This is equal to the determinant of the same algebra in polar form:

$$\det\left(\begin{bmatrix} e^a & 0 & 0 \\ 0 & e^a & 0 \\ 0 & 0 & e^a \end{bmatrix}\begin{bmatrix} v_A(b,c) & v_B(b,c) & v_C(b,c) \\ v_C(b,c) & v_A(b,c) & v_B(b,c) \\ v_B(b,c) & v_C(b,c) & v_A(b,c) \end{bmatrix}\right) = r^3 \qquad (2.2)$$

The rotation matrix has determinant equal to unity, as do all rotation matrices. We have taken $r = e^a$.

Leading to the distance function:

$$r^3 = a^3 + b^3 + c^3 - 3abc \qquad (2.3)$$

Rotation in a spinor space:

The nature of a spinor space is that it has a single rotation expressed as the rotation matrix in the polar form of the algebra. This rotation is of a N-dimensional nature where N is the order of the underlying finite group and the dimension of the spinor space. Spinor rotation is always rotation about a point and not rotation about an axis or more than one axis.

Our experience of our 4-dimensional space-time has accustomed us to only 2-dimensional rotations, and so we find the idea of a 3-dimensional rotation or a 4-dimensional rotation to be strange. The essence of these higher dimensional rotations is that the spinor spaces derive from the finite groups. The finite groups have a clear, and well understood, sub-group structure; for example, the order three group C_3 does not have an order two sub-group. This means that the 3-dimensional spinor algebras[15] do not have 2-dimensional sub-algebras, and this in turn means that the 3-dimensional rotation matrix of the polar form of the 3-dimensional spinor algebras does not have a 2-dimensional rotation matrix within it. We cannot reduce a 3-dimensional rotation into a 2-dimensional rotation.

Of course, some groups, like C_4, do have an order two sub-group. We might expect to find a 2-dimensional rotation as a sub-rotation the C_4 4-dimensional rotation. We do find a kind of 2-dimensional rotation within the 4-dimensional C_4 rotation, but it is a double cover rotation. An example of such a 2-dimensional double cover rotation within a 4-dimensional rotation is:

$$
C_{4|c=d=0} \sim \begin{bmatrix} \cos\theta & \sin\theta & 0 & 0 \\ -\sin\theta & \cos\theta & 0 & 0 \\ 0 & 0 & \cos\theta & -\sin\theta \\ 0 & 0 & \sin\theta & \cos\theta \end{bmatrix} \quad (2.4)
$$

We have rotation in both clockwise and anti-clockwise directions at the same time – rather symmetrical.

Underlying the above 2-dimensional double cover rotation, (2.4), is the fact that setting the other two variables to zero is merely a change of co-ordinate system. A mere change of co-ordinate system cannot change the basic nature of something like a rotation. If we take the eigenvectors of the above '2-dimensional' rotation matrix, (2.4), we get no eigenvectors which are independent of the rotation angle variable – there are no 1's on the leading diagonal. Thus, this '2-dimensional' double cover rotation is really a 4-dimensional rotation written in a

---

[15] We use the terms 'spinor space', spinor algebra', 'division algebra', 'type of complex numbers' interchangeably. These terms are just different names for the same thing.

'special' co-ordinate system. It is rotation about a point and not rotation about an axis.

The 2-dimensional sub-algebra of the 4-dimensional C_4 algebra is a separate mathematical entity, and it is algebraically isomorphic to one of the 2-dimensional spinor algebras, the complex numbers, $\mathbb{C}$, or the hyperbolic complex numbers, $\mathbb{S}$, but it is not identical to either of these algebras – it is a double cover of one of these algebras. Algebraic isomorphism is not identity.

The C_3 spaces:

Every finite group contains spinor spaces. For example the group C_4 contains two different types of spinor spaces, the E-type and the H-type. Most often a particular finite group will contain more than one copy of a particular algebra. The group C_4 contains four E-type spinor spaces and four H-type spinor spaces[16]. The E-type spinor spaces of the group C_4 are algebraically isomorphic to each other – they are the same spinor algebra – but they are written in different bases. Similarly, the H-type spinor spaces of the group C_4 are algebraically isomorphic to each other – they are the same spinor algebra – but they are written in different bases.

The order eight di-cyclic group, which is also misleadingly called the quaternion group, holds only one spinor space, but it holds 128 copies of this spinor space.

We will take the group C_3 as an example. This group holds four spinor spaces:

---

[16] See : Dennis Morris : The Uniqueness of our Space-time.

$$\exp\left(\begin{bmatrix} a & b & c \\ c & a & b \\ b & c & a \end{bmatrix}\right), \quad \exp\left(\begin{bmatrix} a & b & c \\ c & a & -b \\ b & -c & a \end{bmatrix}\right)$$

$$\exp\left(\begin{bmatrix} a & b & c \\ -c & a & b \\ -b & -c & a \end{bmatrix}\right), \quad \exp\left(\begin{bmatrix} a & b & c \\ -c & a & -b \\ -b & c & a \end{bmatrix}\right)$$

(2.5)

These correspond to, in non-matrix notation[17]:

$$a + b\sqrt[3]{+1} + c\sqrt[3]{+1}$$
$$a + b\sqrt[3]{-1} + c\sqrt[3]{+1}$$
$$a + b\sqrt[3]{+1} + c\sqrt[3]{-1}$$
$$a + b\sqrt[3]{-1} + c\sqrt[3]{-1}$$

(2.6)

We see that the middle two algebras are the same algebra. We thus have three algebraically non-isomorphic algebras within the group C_3, but we have two copies of one of these algebras.

This algebraic isomorphism is apparent in the form of the distance functions of these spaces. We have, respectively:

$$d^3 = a^3 + b^3 + c^3 - 3abc$$
$$\begin{cases} d^3 = a^3 + b^3 - c^3 - 3abc \\ d^3 = a^3 - b^3 + c^3 - 3abc \end{cases}$$
$$d^3 = a^3 - b^3 - c^3 - 3abc$$

(2.7)

The $C_2 \times C_2$ spaces:

The commutative group $C_2 \times C_2$ contains sixteen spinor spaces; eight of these spinor spaces are commutative division algebras, and,

---

[17] We have to be a little cautious with non-matrix notation because only the polar form of the algebra is a division algebra, and, when we write the algebra without the exponential, we are not really presenting a proper algebra.

remarkably, eight of these spinor spaces are non-commutative division algebras. The eight non-commutative spinor spaces are of great interest to ourselves.

Within the $C_2 \times C_2$ group, there are two copies of the quaternion algebras, the left-chiral quaternions and the right-chiral quaternions. Within $C_2 \times C_2$, there are six copies of the A_3 algebras, three left-chiral forms and three right-chiral forms.

The distance functions of the eight non-commutative $C_2 \times C_2$ algebras are the two quaternions:

$$d^2 = t^2 + x^2 + y^2 + z^2$$
$$d^2 = t^2 + x^2 + y^2 + z^2$$

(2.8)

And the six A_3 algebras:

$$d^2 = t^2 + x^2 - y^2 - z^2$$
$$d^2 = t^2 + x^2 - y^2 - z^2$$
$$d^2 = t^2 - x^2 - y^2 + z^2$$
$$d^2 = t^2 - x^2 - y^2 + z^2$$
$$d^2 = t^2 - x^2 + y^2 - z^2$$
$$d^2 = t^2 - x^2 + y^2 - z^2$$

(2.9)

The emergent expectation spaces of a finite group:
We take every spinor space of an isomorphic set, for example we would take both the quaternion spinor spaces above, (2.8), and we superimpose those spaces. Basically, we just pile the isomorphic spaces on top of each other. What results is an emergent expectation space. We are adding isomorphic copies of algebras that are written in different bases.

The distance functions of the isomorphic spaces are added; for example, we add the six distance functions of the six A_3 algebras, (2.9). This gives the emergent expectation distance function. This is the distance function of the emergent expectation space. The distance function of the A_3 emergent expectation space is:

$$6d^2 = 6t^2 - 2x^2 - 2y^2 - 2z^2 \qquad (2.10)$$

We can tidy this a little by dividing by two:

$$3d^2 = 3t^2 - x^2 - y^2 - z^2 \qquad (2.11)$$

We can ignore the three in front of the d^2 because this is no more than a scaling factor, and we can be rid of the three in front of the t^2 by adjusting the units in which we measure time. This gives:

$$d^2 = t^2 - x^2 - y^2 - z^2 \qquad (2.12)$$

This is the distance function of our 4-dimensional space-time.

The dimension of the emergent expectation space is the same as the dimension of the spinor spaces which were superimposed to form it because this is the number of variables.

The emergent expectation space has as many angles as there are superimposed spinor algebras – one from each spinor algebra. In the case of the A_3 algebras, there are six A_3 algebras and so there are six angles in the A_3 emergent expectation space.

Coincidently, if there is such a thing as coincidence in mathematics, the A_3 emergent expectation distance function is such that it will support six 2-dimensional rotations – there are six different pairs of variables which form a quadratic form within the distance function. This remarkable coincidence is unique within all the infinitely many finite groups and their spinor spaces. This is our unique 4-dimensional space-time.

Consequences of superimposition:

Adding algebras written in different bases destroys the algebras. We are left with the debris of that destruction. That debris is four variables. These four variables cannot be imaginary variables because, with no algebra, there is no multiplication, and so we cannot have relations like $i^2 = -1$. The four debris variables are all real variables; we have $\mathbb{R}^4$.

Every spinor algebra is flat; indeed, it is so flat that it does not even have zero curvature. The algebra falls to bits if we try to impose some kind of curvature on to a spinor space. With the destruction of the algebras comes the destruction of any kind of flatness other than at an infinitesimally small point[18] – there is no affine connection within the space, to be technical. It would seem as though the emergent expectation space is a space with no sense of parallel lines – no affine structure, but this is not so.

The six A_3 angles are added to form a concept of angle from point to point in the emergent expectation space. Acting as a gauge space, the A_3 angle varies locally (from point to point) in the emergent expectation space, and this imposes on to the emergent expectation space an affine connection[19].

Normally, within a fibre bundle of spaces, the locally varying phase of a gauge space causes an 'antagonism' between the affine connection of the underlying space, which is always 4-dimensional space-time, and the affine connection of the gauge space. This is expressed as a force like the electromagnetic force or the weak force or the strong force. In the absence of an affine connection in the underlying space, the gauge affine connection simple becomes the affine connection of the underlying space. Thus, the emergent expectation space acquires an affine connection – a sense of parallel transport from point to point within it. We have our curved 4-dimensional space-time.

---

[18] At such a point, we can fit a flat tangent space or a fabrication of flat tangent spaces. Which we fit is determined by the emergent distance function – see Dennis Morris : Upon General Relativity.

[19] This is dealt with in mathematical detail within the book : Dennis Morris : Upon General Relativity.

One of the consequences of this lack of antagonism between the affine connection of the A_3 emergent expectation space and an underlying 'base space' in a fibre bundle is that there is no gauge boson. This, we opine, means there is no graviton.

The gauge spaces of particle physics which are attached to our 4-dimensional space-time in a fibre bundle each have gauge bosons associated with them. For the $U(1)$ gauge space, the gauge boson is the photon of the electromagnetic force. The $SU(2)$ gauge space has three gauge bosons associated with it; these are the $\{W^{\pm}, Z^0\}$ of the weak force. The conventional $SU(3)$ gauge space has eight gauge bosons associated with it; these are the eight gluons of the strong force.

Chapter 3

Why our Space-time is Unique

Each spinor space exists in its own right with its own single rotation about a point. The spinor rotation is of the same dimension as the order of the group. Each spinor space has its own distance function. For reasons that are still not understood, these spinor spaces do not manifest themselves geometrically. They exist as the real numbers exist without concrete manifestation.

Quite why a superimposition of isomorphic spinor algebras should manifest itself geometrically is also not understood, but observation seems to fit with the idea that it does so manifest itself. So why do we not see 5-dimensional emergent expectation spaces or even the 3-dimensional emergent expectation space?

The answer seems to be that no emergent expectation space other than the A_3 emergent expectation space can fully hold any form of rotation.

The quaternion emergent expectation space can partially hold rotation, and it seems that the antagonism between the quaternion emergent expectation space and our 4-dimensional space-time is what we call the electroweak force. To understand this better, we will take a look at some emergent expectation distance functions.

True rotation:
True rotation exists in only spinor algebras. These are the only mathematical constructions that have trigonometric functions and rotation matrices without zeros. We can begin with the 2-dimensional spinor algebras and write down the distance functions respected by the different types of rotations. We have done this for 2-dimensions above, (1.5) & (1.6).

We can then pass on to the 3-dimensional spinor spaces and similarly write down the distance functions respected by the 3-dimensional rotations. We have done this above, (2.7). Similarly we can collect together all the distance functions respected by all the different types of rotations of all dimensionality which emerge from every finite group. We now have the distance functions respected by every possible rotation.

Having collated our list of distance functions respected by all the different types of rotation, we then form the emergent expectation distance functions of all the sets of isomorphic algebras from all the finite groups and we ask which of these emergent expectation distance functions can hold any type of rotation[20]. By this we mean which of these emergent expectation distance functions will reduce to a distance function respected by a spinor rotation in a way similar to how the expectation distance function of the emergent A_3 expectation space, (2.12), reduces to distance functions respected by the 2-dimensional rotations, (1.8). The answer is only the A_3 emergent expectation distance function can hold a full set of rotations. Our 4-dimensional space-time is unique.

We have above, (1.12), calculated the only 3-dimensional emergent distance function. Clearly, this does not have the form respected by either the 3-dimensional rotations or by the 2-dimensional rotations. Here we have an example of what we will find many times (an infinite number of times actually) when we calculate an emergent expectation distance function.

In general, the distance function of a N-dimensional spinor algebra, that is the determinant of the matrix form, will be of power N. The polar form of a spinor algebra includes a $n \times n$ radial matrix which is of the form:

---

[20] This is done in : Dennis Morris : The Uniqueness of our Space-time.

$$
\begin{bmatrix}
r & 0 & 0 & . & . & . \\
0 & r & 0 & . & . & . \\
0 & 0 & r & . & . & . \\
. & . & . & . & . & . \\
. & . & . & . & . & . \\
. & . & . & . & . & .
\end{bmatrix}
\tag{3.1}
$$

This will always have a determinant of the form r^n. The rotation matrix will always have determinant equal to unity; and so the determinant, distance function, of a spinor space will always be to the power n. Thus, the distance functions respected by the spinor rotations will always be to power n; 2-dimensional rotations will respect quadratic distance functions; 3-dimensional rotations will respect cubic distance functions etc..

Consider the emergent expectation distance function of a set of isomorphic spinor algebras of dimension p where p is a prime number. Such an emergent expectation distance function can never be factored into lesser powers, and so it can never respect lesser dimensional rotations. In a stroke, we have eliminated all groups of prime order from our list of possible fully rotational emergent expectation spaces.

Now consider a group which has a sub-group of prime order other than C_2. The prime order sub-group prevents factorisation of the emergent expectation distance function into a function which avoids the prime sub-group order. Such groups cannot support an emergent expectation space with full rotation. Further, when we have established that a particular group does not support a fully rotational emergent expectation space, D_4 is such a group, then we have established that any group which has this group as a sub-group does not support a fully rotational emergent expectation space[21].

---

[21] Not quite! There are groups of order p^n to consider.

Although we will not go into details, no cyclic group other than C_2 can support a fully rotational emergent expectation space.

The reader will see that our infinite list of groups which might support a fully rotational emergent expectation space grows thin.

We then need the correct number of angles to match the number of sub-spaces which can be formed from the number of axes. The reader will recall that there are six pairs of axes in 4-dimensional space and there are six A_3 algebras.

Detailed analysis leads to the view that only the A_3 emergent expectation space can be manifest as a fully rotational geometric space, but the proof of this is not yet complete[22].

An intriguing matter:

Consider a group like C_4. It holds the two types of C_4 trigonometric functions. These two types of trigonometric functions differ from each other by no more than the position of minus signs in the spinor matrix in a way similar to how the 2-dimensional Euclidean trigonometric functions are related to the hyperbolic trigonometric functions:

$$\cosh(ix) = \cos x$$
$$\sinh(ix) = i \sin x \tag{3.2}$$

The group C_4 has a C_2 sub-group. The C_2 sub-group holds the 2-dimensional trigonometric functions, and so we know that the C_4 trigonometric functions must reduce to the C_2 trigonometric functions when two of the 4-dimensional spinor variables are zero. This very much restricts the form which the C_4 trigonometric functions might take. Now consider the order twelve group A_4. This group has two C_2 sub-groups, four C_3 sub-groups, and a $C_2 \times C_2$ sub-group. How would you design a 12-dimensional trigonometric function that could reduce

---

[22] See : Dennis Morris : The Uniqueness of our Space-time.

to all the different trigonometric functions implied by the presence of all the sub-groups of A_4? It's miraculous stuff, but the mathematics does this.

What we have said about the trigonometric functions also applies to the distance functions. What kind of 4-dimensional distance function will reduce to 2-dimensional distance functions when two of the variables are zero? One such distance function springs to mind; it is a quadratic form:

$$d^4 = \left(t^2 + x^2 - y^2 - z^2\right)^2 \tag{3.3}$$

There is also the C_4 distance function:

$$d^4 = \left((a+c)^2 - (b+d)^2\right)\left((a-c)^2 + (b-d)^2\right)$$
$$d^2\big|_{b=d=0} = a^2 - c^2 \tag{3.4}$$

Thus this C_4 space will reduce to a 2-dimensional rotation, but the associated emergent expectation space will not so reduce.

The intriguing bit:
Interestingly, there is an 8-dimensional set of isomorphic algebras from the $C_2 \times C_2 \times C_2$ group which does have an emergent expectation space that will reduce to our 4-dimensional space-time for particular values of the eight variables. We think this is the strong nuclear force, but our thought is little more than speculation at present.

Chapter 4

The Expanding Universe and Dark Energy

For almost a century since Edwin Hubble (1889-1953) and Milton L. Humason (1891-1972) discovered the expansion of our universe, humanity has been struck speechless by the observation that empty space is expanding. How can empty space get emptier?

Looking at the spinor algebras, it seems that it is entirely natural for empty space to expand. There are only three spinor spaces which do not expand; these are the Euclidean complex plane and the two quaternion spaces. We should be speechless at the existence of three spaces which do not expand. The majority of spinor spaces expand for the same reason that our 4-dimensional space-time expands. The expansion is intrinsic to the distance function.

Our universe:

The distance function of our 4-dimensional space-time is:

$$dist^2 = t^2 - x^2 - y^2 - z^2 \qquad (4.1)$$

For simplicity, we ignore two of the three spatial dimensions and we have a hyperbola:

$$dist^2 = t^2 - y^2 \qquad (4.2)$$

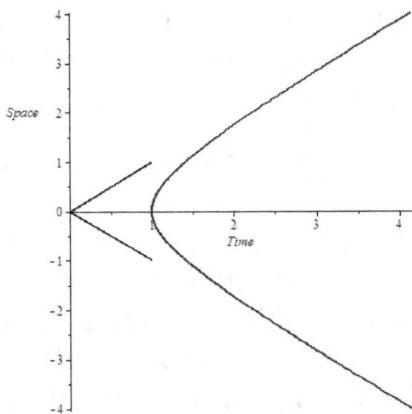

Within this plot, the speed of light is the 45^0 asymptotes; the axes are scaled differently to ease the presentation. The hyperbola is the maximum extent of space. All of space is between the two limbs of the hyperbola. We see this maximum extent of space increases as time increases. We see the rate of expansion approaches the asymptotic speed of light as time increases but that the rate of expansion is always greater than the speed of light. At $time = 1$, we have an infinite rate of expansion for zero time. Of course, in 4-dimensional space-time, the hyperbola is 4-dimensional, but that is hard to draw.

We assert that this is the expanding universe. We are asserting that the expansion of the universe is within the distance function, and that it is entirely natural for a space with the distance function (4.1) to expand as time increases. Of course, we might have put 'age' on the horizontal axis rather than 'time' and 'distance to the edge of the observable universe' rather than 'space' on the vertical axis. We have an inflationary beginning to the universe at time $t = 1$, and we have the universe approaching the 'flat' asymptote of expanding forever but at an always slowing rate of expansion. We have no need of 'phase changes' to initiate the inflationary start nor of gravitation to slow the expansion.

For many decades, cosmologists took the view that the universe began as a huge explosion, called the big bang, which threw the matter of the universe out into a pre-existing empty space rather like a grenade exploding in a room. Only lately have cosmologists come to realise that the expanding universe is nothing like an exploding grenade. The modern view is that space and time were not pre-existing but that they began to unfold expansively 13.8 billion years ago.

Within the exploding grenade understanding, there is the expectation that gravity will slow the rate of expansion as gravity would slow the velocity of a piece of grenade shrapnel as it flew upward from an exploding grenade. This leads to the question of whether or not the expansion of the universe will reach escape velocity as a piece of grenade shrapnel might reach escape velocity. Observations in the second half of the 20th century concluded that the universe had exactly the right amount of mass to match escape velocity; that is not enough mass to reverse the expansion but enough mass to stop escape. Prior to this realisation, cosmologists entertained the idea that the universe might one day start to contract as a piece of grenade shrapnel might eventually stop rising and begin to fall back toward the Earth.

These scenarios of possible contraction or eternal escape were based upon the idea that gravity slows expansion. If the expansion of the universe is within the distance function, as we assert it is, then gravity plays no role in slowing the expansion, and so the amount of mass in the universe is irrelevant to the rate of expansion.

Well, that is all very neat. As soon as we overcome our prejudice that empty space does not expand, everything fits together; but does this neat scenario fit observation?

Dark energy:
Recent observations of 1a type supernovae have caused consternation and led to the view that the expansion of the universe seems to be accelerating of late. We are given the idea that the expansion of the universe began with an inflationary start and the expansion then started to slow, as expansion in a properly behaved universe would. However, about nine billion years ago, something, now called dark energy, took control of the universe and began to accelerate the expansion. Well, that seems to put the kibosh upon our assertion that the expansion of the universe is inherent within the distance function. There is reality, and

there is seeming reality. Let us firstly examine the observations, and let us secondly examine the observers.

Type 1a supernova are such that they have a standard brightness. This is well understood, and no-one questions it. Since these supernova have a standard brightness they can be used as standard candles to measure distance. That is a measure of distance. Since distance equates to looking back in time, we see these supernovae as they were billions of years ago.

As with all distant bodies, the type 1a supernova have a redshift which is a measure of how fast they are moving away from ourselves. We now have a measure of the rate of expansion. Putting together the distance measurements and the rate of expansion measurements, we have a knowledge of how the rate of expansion has changed over the history of the universe.

This newly gained knowledge of expansion rates seems to conflict with our assertion about the expansion of the universe being intrinsic to the distance function of the universe; however, this newly gained knowledge is based upon assumptions about redshift which our assertion calls into question. We do not question the observations. We question the position of the observers.

Let us consider a modern observer looking at the universe. As the observer looks toward distant galaxies, she looks back in time. On the space-time diagram below, she looks back at the speed of light which is 45^0 to the axis. We have:

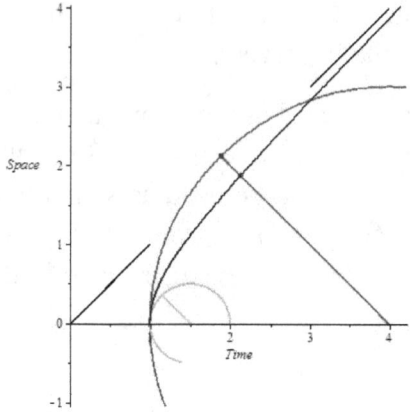

An observer looking at the universe from, say, 3-'billion' years old (the point 4), is able to see 'outside' of the universe (beyond the hyperbola) because she is able to see 3-billion years into the past. Hang on!, you cannot see outside of the universe. The observer believes she is seeing 3-billion years into the past, but the edge of the universe is less than 3-billion light-years away. How does this affect the redshift and the brightness measurements?

The situation is different for an observer looking at the universe from, say, a half a billion years old. In this case, we have:

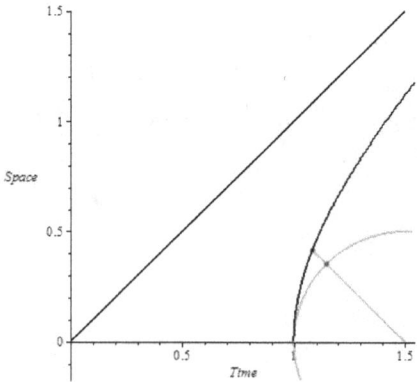

In this case, the observer looks back through half a billion years and to half a billion light-years distant, but she does not see to the edge of the universe. Clearly, she might think she is seeing right back to the big bang because the big bang was half a billion years earlier and she is looking back through half a billion years. There might be discrepancy between her redshift measurements and brightness measurements.

How does the discrepancy between the two distances of furthest extent of observation and the distance to the edge of the universe vary since the start of the universe? We have, with unfair but suggestive labels on the axes:

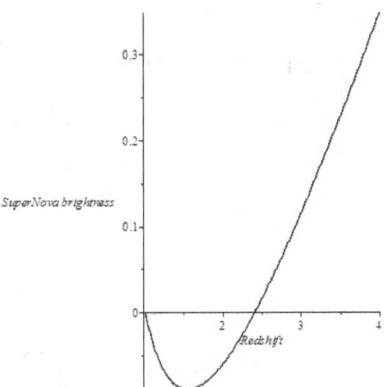

The calculations of this are presented later.

An understanding might be that, in the early universe, the supernovae are farther away, on the hyperbola, than we think them to be by their redshift, and so they are fainter. In the later universe, the supernovae are closer than we think them to be by their redshift, and so they are brighter.

Alternatively, for the age of the universe, the speed of light might have been a constant:

$$c = \frac{\text{distance to edge of universe}}{\text{age of universe}} \tag{4.3}$$

If this were the case, we would expect discrepancies in the redshift measurements.

Perhaps, as well as expansion being intrinsic to the distance function of the universe, apparent dark energy is also intrinsic to that distance function.

Calculations:

For a given point T on the time axis (this is time since the big bang plus one), an observer at T will see $(T-1)$ light-years into the past (circles) to the point:

$$y = (T-1)\sin\left(\frac{\pi}{4}\right) \qquad x = T - (T-1)\cos\left(\frac{\pi}{4}\right) \qquad (4.4)$$

However, the expanding universe is at the point in the observer's line of sight given by the intersection of the hyperbola, $y = \sqrt[3]{x^2 - 1}$ and the line $y = -(x-T)$. We have:

$$y^2 = x^2 - 1 = x^2 + T^2 - 2xT \qquad \Rightarrow x = \frac{T}{2} + \frac{1}{2T}$$

$$y = -(x-T) = -\left(\frac{1}{2T} + \frac{T}{2} - T\right) \qquad \Rightarrow y = \frac{T}{2} - \frac{1}{2T} \qquad (4.5)$$

The difference in the co-ordinates of the points on the circles and the points on the hyperbola is:

$$y_{hyp} - y_{circ} = \frac{T}{2} - \frac{1}{2T} - (T-1)\sin\left(\frac{\pi}{4}\right) = \frac{T^2-1}{2T} - \frac{(T-1)}{\sqrt[3]{2}} \qquad (4.6)$$

$$x_{hyp} - x_{circ} = \frac{T}{2} + \frac{1}{2T} - \left(T - (T-1)\cos\left(\frac{\pi}{4}\right)\right)$$

$$= \frac{T^2+1}{2T} - T + \frac{(T-1)}{\sqrt[3]{2}} \qquad (4.7)$$

$$= \frac{1-T^2}{2T} + \frac{T-1}{\sqrt[3]{2}}$$

The actual distance between the points is (Pythagoras):

$$dist^2 = \left(\frac{T^2-1}{2T} - \frac{(T-1)}{\sqrt[2]{2}} \right)^2 + \left(\frac{1-T^2}{2T} + \frac{(T-1)}{\sqrt[2]{2}} \right)^2$$

$$dist = \frac{1}{\sqrt[2]{2}} \frac{(T-1)(\sqrt[2]{2}T - T - 1)}{T}$$

(4.8)

This is the graph given above.

Chapter 5

The Gauge Spaces

There is a quaternion emergent expectation space which has two angles of rotation within it reflecting its derivation from the two quaternion spinor spaces. The emergent expectation distance function of the quaternion space, see (1.9), is such that it coincides with the distance function of our 4-dimensional space-time, (1.7), in two cases.

The first case is when all the spatial variables are zero. The distance functions of the emergent quaternion space and our 4-dimensional space-time then reduce to simply the time variable:

$$d^2_{Space-time} = t^2 \qquad\qquad d^2_{Quaternion} = t^2 \qquad\qquad (5.1)$$

This effectively says that, these two spaces will coincide provided the coincidence is of zero spatial extent – a point particle like the electron. The coincidence will endure through time.

The second case is when the time variable is zero. In this case, the distance functions of the emergent quaternion space and our 4-dimensional space-time then reduce to:

$$d^2_{Space-time} = -x^2 - y^2 - z^2 \qquad\qquad d^2_{Quaternion} = x^2 + y^2 + z^2 \qquad (5.2)$$

These are the same except for the meaningless sign. This effectively says that, these two spaces will coincide provided the coincidence is of zero temporal extent. The coincidence can have spatial extent but only instantaneously – like an electron passing through two spatially separated slits.

The 2-dimensional emergent expectation spaces are just the spinor spaces because there is only one copy of each of them. Clearly, the distance functions of the 2-dimensional spaces coincide with the distance function of our 4-dimensional space-time.

It seems that this coincidence of distance functions ties different spaces together. It seems that we might have a fibre bundle in which our 4-dimensional space-time is the underlying space and the quaternion emergent space and the two 2-dimensional spaces are fixed, at each point in our 4-dimensional space-time in a way similar to how Newton envisaged 3-dimensional space to be fixed to time.

The fibre bundle view of space is called gauge theory. Gauge theory envisions three gauge spaces called $U(1)$, which is the complex plane, $\mathbb{C}$, called $SU(2)$, which is quaternion space, and called $SU(3)$, of which we have made no mention so far. We think $SU(3)$ does not really exist and that the physicists have got this bit wrong.

Gauge theory does not propose any explanation of how these different spaces are fixed together. The proposal that these spaces are fitted together by the coincidence of distance functions is the child of your author and might be completely wrong.

Within gauge theory, the forces of nature other than gravitation arise as an antagonism between the affine connections of the gauge spaces and the underlying 4-dimensional space-time. This is referred to as locally varying phase – which rather fogs what is really happening. Along with the antagonism between the affine connections comes a covariant derivative which has an extra term which corresponds to a potential. The potential is electromagnetic in the case of $U(1)$, the complex plane, and is the weak force in the case of $SU(2)$, the quaternions, and is the strong force in the case of $SU(3)$, which we think ought to be an 8-dimensional algebra from the $C_2 \times C_2 \times C_2$ algebras.

Other Books by the Same Author

The Naked Spinor – a Rewrite of Clifford Algebra

Spinors exist in Clifford algebras. In this book, we explore the nature of spinors. This book is an excellent introduction to Clifford algebra.

Complex Numbers The Higher Dimensional Forms – Spinor Algebra

In this book, we explore the higher dimensional forms of complex numbers. These higher dimensional forms are closely connected to spinors.

Upon General Relativity

In this book, we see how 4-dimensional space-time, gravity, and electromagnetism emerge from the spinor algebras. This is an excellent and easy paced introduction to general relativity.

From Where Comes the Universe

This is a guide for the lay person to the physics of empty space. We seek to avoid complicated mathematics within this short volume.

Empty Space is Amazing Stuff – The Special Theory of Relativity

This book deduces the theory of special relativity from the finite groups. It gives a unique insight into the nature of the 2-dimensional space-time of special relativity. The whole of the theory is introduced including the acceleration transforms which are absent from almost all other treatments of this area of physics.

The Nuts and Bolts of Quantum Mechanics

This is a gentle introduction to quantum mechanics for undergraduates.

Quaternions

This book pulls together the often separate properties of the quaternions. Non-commutative differentiation is covered as is non-commutative rotation and non-commutative inner products along with the quaternion trigonometric functions.

The Uniqueness of our Space-time

This book reports the finding that the only two geometric spaces within the finite groups are the two spaces which together form our universe. This is a startling finding. The nature of geometric space is explained alongside the nature of division algebra space. This book is a catalogue of the higher dimensional complex numbers up to dimension fifteen. The computer code is included.

Lie Groups and Lie Algebras

This book presents Lie theory from a diametrically different perspective to the usual presentation. This makes the subject much more intuitively obvious and easier to learn. Included is perhaps the clearest and simplest presentation of the true nature of the Lie group $SU(2)$ ever presented.

The Physics of Empty Space

This book presents a comprehensive understanding of empty space. The presence of 2-dimensional rotations in our 4-dimensional space-time is explained. Also included is a very gentle introduction to non-commutative differentiation. Classical electromagetism is deduced from the quaternions.

The Electron

This book presents the quantum field theory view of the electron and the neutrino. This view is radically different from the classical view of the electron presented in most schools and colleges. This book gives a very clear exposition of the Dirac equation including the quaternion rewrite of the Dirac equation. This is an excellent introduction to particle physics for students prior to university, during university and after university courses in physics.

The Quaternion Dirac Equation

This short book derives a quaternion form of the Dirac equation. The convention Dirac equation is set in a 16-dimensional Clifford algebra. The 4-dimensional quaternion form of the Dirac equation is much simpler and mathematically more straight-forward than the conventional Dirac equation. The quaternion Dirac equation leads to a non-chiral (spin either up or down) massive electron field and to a massless neutrino field. The massless neutrino field is chiral leading to only left-chiral neutrinos. Although the neutrino field is massless, allowing neutrinos to travel at the speed of light, the neutrino field squared is massive allowing neutrino oscillation.

An Essay on the Nature of Space-time

This small and inexpensive volume presents a view of the nature of empty space without the detailed mathematics. The expanding universe and dark energy is discussed.

Elementary Calculus from an Advanced Standpoint

This book rewrites the calculus of the complex numbers in a way that covers all division algebras and makes all continuous complex functions differentiable and integrable. Non-commutative differentiation is covered. Gauge covariant differentiation is covered as is the covariant derivative of general relativity.

Even Mathematicians and Physicists make Mistakes

This book points out what seems to be several important errors of modern physics and modern mathematics. Errors like the misunderstanding of rotation, the failure to teach the higher dimensional complex numbers in most universities, and the mathematical inconsistency of the Dirac equation and some casual errors are discussed. These errors are set in their historical circumstances and there is discussion about why they happened and the consequences of their happening. There is also an interesting chapter on the nature of mathematical proof within our society, and several famous proofs are discussed (without the details).

Finite Groups – A Simple Introduction

This book introduces the reader to finite group theory. Many introductory books on finite groups bury the reader in geometrical examples or in other types of groups and lose the central nature of a finite group. This book sticks firmly with the permutation nature of finite groups and elucidates that nature by the extensive use of permutation matrices. Permutation matrices simplify the subject considerably. This book is probably unique in its use of permutation matrices and therefore unique in its simplicity.

The Left-handed Spinor

This book covers the left-handed parts of mathematics which we call the chiral algebras. These algebras have CP invariance, violation of parity, and many other aspects which makes them relevant to theoretical physics. It is quite a revelation to discover that mathematics is left-handed.

Non-commutative Differentiation and the Commutator

(The Search for the Fermion Content of the Universe)

This book develops the theory of non-commutative differentiation from the fundamentals of algebra. We see what an algebraic operation (addition, multiplication) really is, and we discover that the commutator is a third fundamental algebraic operation within some division algebras. This leads to the first part of the derivation of the fermion content of the universe.

Index

Index

S

sets of permutations, 16
space-time plane, 4
space-time rotation, 6
special relativity, 4
spinning disc, 8
spinor algebra, 14
spinor algebras, 16
spinor rotation, 7, 17, 25
spinor space, 8
spinor spaces, 16
standard candles, 33
strong force, 23, 24, 39
SU(2), 39
SU(3), 39
sub-algebra, 18
sub-group, 18
supernova, 32

T

trigonometric function, 14
trigonometric functions, 28
true rotation, 25

U

U(1), 39
unique 4-dimensional space-time, 22
uniqueness, of our space-time, 15
unity of space and time, 5
universe, expanding, 31

W

weak force, 23, 24, 39
Weyl spinor, 16

www.ingramcontent.com/pod-product-compliance
Lightning Source LLC
Chambersburg PA
CBHW071828200526
45169CB00018B/1232